# (R.E.A.L)

## Reconciled Ethereal Augmentations of Logic

*(R.E.A.L.)*
*Reconciled Ethereal Augmentations of Logic*
Copyright © 2023 by Bruce Robert Nye Jr.



Paperback ISBN: 979-8-8229-1702-6
eBook ISBN: 979-8-8229-1703-3

# (R.E.A.L)

## Reconciled Ethereal Augmentations of Logic

Revealing Planetary Spacing, Habitable Exoplanet Numeric Identification, and Explaining Gravitation

**BRUCE ROBERT NYE JR.**

Dedicated to Paul Franklin Nye, Jr., my grandfather
and mentor. You always respected and reared my
unabashed thirst for knowledge. Such a tidal wave of
love and compassion crests once in a soul's lifetime.

# TABLE OF CONTENTS

# INTRODUCTION

This work aims to create a non prevaricatory and simplistic method of understanding the physical heavens that border humankind from the infinite, and a means to understand nearly all the motion in the universe. A work so pure in reasoning and constituents that even the laymen with no particular specialized background could invoke and utilize the productions. Meaning simple explanations are inversely proportional to the complexity and length of the syntax describing the concept in question. A simplistic approach (to any problem) always shares a greater continuity with understanding. This truth doesn't simply act as the conduit for sentient understanding of our cosmos, but also an abridgment to the empirical logic governing the physical universe. Constructing a system of mathematical relation that employs logic as the Rosetta Stone interpretation. Creating a language of understanding concerning natural mechanisms, and reconciling concepts believed to be fact.

This work's overall construction of data delivery, with respect to the information at hand, is aimed simply at classic logical interpretation. This is done by the simple means of classification of logistics, meaning the subject of any instance must be constructed with the elevation of thought to be considered logical. A binary sense of reality will often deliver the most relatable results. And in that spirit, this work is structured to acutely arouse the logical sense of the reader. The proposed manuscript is being presented in stages. The book is divided into three parts, each of which collectively unveils a portrait of harmonious divinity.

Part One begins with the challenge of the accepted Titius–Bode law (sometimes termed just Bode's law), which is a formulaic prediction of spacing between planets in any given solar system. The formula suggests that (extending outward) each planet should be approximately twice as far from the sun as the one before. The hypothesis correctly anticipated the orbits of Ceres (in the asteroid belt) and Uranus but failed as a predictor of Neptune's orbit. It is named after Johann Daniel Titius and Johann Elert Bode [ref https://www.britannica.com/science/Bodes-law]. An argument is presented with data collected from NASA's Kepler probe over the last fifteen years. Clear evidence will be presented, showing a decisive pattern connected to a geometric number sequence! This sequence not only exposes a strong Fibonacci relationship, but is dictating planetary spacing and incidentally predicts a clear "bound" with respect to average planetary concentration per system. The inadvertent way this understanding is interpreted is simply by the fact that the sequence in question is finite, with only eleven terms! This is a clear indication of allowance with respect to planetary count per system, as well as a clear bound of average orbital distance. As will be shown below, star systems like our own are entirely unique. The majority of systems (nearly 2,300) have a single orbital companion. Then two companions, respectively, three companions, and so on to our own system and Kepler-90, which both have eight orbital companions.

Orbital patterns within star systems (of all magnitudes) are a long-documented phenomenon. Observations have shown that exoplanets spaced in ratios that are evenly expressed integers have more stable orbital durations, eccentricities, and overall consistency. The sequence itself is describing a pattern long recognized as orbital resonance. In celestial mechanics, orbital resonance occurs when orbiting bodies exert regular, periodic gravitational influence on each other, usually because their orbital periods are related by a ratio of small integers. Most commonly this relationship is found between a pair of objects (binary resonance). The physical principle behind orbital resonance is similar in concept to pushing a child on a swing, whereby the orbit and the swing both have a natural frequency, and the body doing the "pushing" will act in periodic repetition to have a cumulative effect on the motion. Orbital resonances greatly enhance the mutual gravitational influence of the bodies (i.e., their ability to alter or constrain each other's orbits). In most cases this results in an unstable interaction in which the exoplanets exchange momentum and shift orbits until the resonance no longer exists. Under some circumstances a resonant system can be self-correcting and thus stable. Examples are the 1:2:4 resonance of Jupiter's moons Ganymede, Europa, and Io, and the 2:3 resonance between Pluto and Neptune. Unstable resonances with Saturn's inner moons give rise to gaps in the rings of Saturn. The special case of 1:1 resonance between bodies with similar orbital radii causes large solar system bodies to eject most other bodies sharing their orbits; this is part of the much more extensive process of clearing the neighborhood, an effect that is used in the current definition of a planet.1

A binary resonance ratio in this article should be interpreted as the ratio of number of orbits completed in the same time interval rather than the ratio of orbital periods, which would be the inverse ratio. Thus the 2:3 ratio above means that Pluto completes two orbits in the time it takes Neptune to complete three. In the case of resonance relationships among three or more bodies, either type of ratio may be used (whereby the smallest whole-integer ratio sequences are not necessarily reversals of each other), and the type of ratio will be specified[ref https://en.wikipedia.org/wiki/Orbital_resonance].

Part Two outlines another numerical phenomenon, only one with a much more deliberate scope. The book takes a turn to an explanation of "habitable zone distribution," *habitable zone*, the orbital region around a star in which an Earth-like planet can possess liquid water on its surface and possibly support life. Liquid water is essential to all life on Earth, and so the definition of a habitable zone is based on the hypothesis that extraterrestrial life would share this requirement[ref https://www.britannica.com/science/habitable-zone].

Using a completely original approach, the semi-major axis of potentially habitable exoplanets is converted to solar radius units (along with orbital duration in Earth years), revealing a stunning numerological correlation. The two planetary variables, respectively, produce Fibonacci integers. Fibonacci integers have been a long-standing mathematical fascination, appearing in botanical geometric patterns, DNA helix, our Milky Way galaxy, etc. Long fantasized as being the inspiration behind Leonardo DaVinci's *Leonardo: The Universal Man*, the irrational integer has had its fair share of fame. A team of researchers in 2012 measured total orbital completions for each planet in our system, and the results were conclusively the Fibonacci sequence. Meaning each planet in our system (going outward) successively produced partitioned Fibonacci integers.

Fibonacci sequence, the sequence of numbers 1, 1, 2, 3, 5, 8, 13, 21, …, each of which, after the second, is the sum of the two previous numbers; that is the nth Fibonacci number, $F_n = F_{n-1} + F_{n-2}$. The sequence was noted by the medieval Italian mathematician Fibonacci (Leonardo Pisano) in his *Liber Abaci* (1202, *Book of the Abacus*), which also popularized Hindu-Arabic numerals and the decimal number system in Europe. Fibonacci introduced the sequence in the context of the problem of how many pairs of rabbits there would be in an enclosed area if every month a pair produced a new pair and rabbit pairs could produce another pair beginning in their second month. The numbers of the sequence occur throughout nature, such as in the spirals of sunflower heads and snail shells. The ratios between successive terms of the sequence tend to the golden ratio $\varphi = (1 + \text{Square root of}\sqrt{5})/2$ or 1.6180 [ref https://www.britannica.com/science/Fibonacci-number].

Any two integers within the Fibonacci sequence(>1) can be considered (a,b), respectively. The same two Fibonacci variables are collected by means of the described method below. It should be noted, not all potentially habitable exoplanets exhibit this particular phenomenon, but all exoplanets that do exhibit the Fibonacci integer extraction are in the habitable zone or are the remnants of the primitive habitable zone before the star(s) of the system in question evolved and deviated the habitable zone outward. A simple quotient can derive an incredibly close approximation of habitable zone divergence based on the solar mass / solar radius divergence.

This highly complex numerical pattern, along with the remarkable number of orbital companions our sun hosts, realistically serves the greatest reason (aside from our proximity to the sun coupled with its apparent luminosity) habitation began and accelerated on our planet. This will also be discussed below in Part Two. Jupiter, beginning itself essentially a small sun, has an enormous gravitation exertion.

The massive gas giant is approximately 2 x 10^27 kg. That's three hundred times the mass of Planet Earth! Jupiter hosts between eighty and ninety-five moons. Some are too small and have such irregular orbits, their status as orbiters is difficult to be certain of. This incredible moon count of Jupiter is a great example of the variant levels of protection our planet receives. The asteroid belt, of course, is another veil of protection. Hosting nearly forty percent of the mass of our moon. Passively orbiting the sun at near-perfect circular orbits. A minefield of ancient rock and ice, hurtling at nearly 45,000 mph. This clear concert of variables has no explanation aside from chance and pure statistical aberration. Which would serve to explain why no other star system of the nearly 5,000 has no sign of bacterial growth, diatomic fluorescence, signs of advanced civilization, etc.

And finally Part Three casually quantifies the force of gravitation and offers a means of propagation, not yet posed, concerning the means in which gravitation manifests and the nature of attraction at the quantum level. Simplicity again is stressed as the conduit of understanding, clear arguments made, and correlating data defending the claim. A question was posed, "What is the effect of gravitation on one particular atom of variant atomic mass?" This portion of the work seeks to answer that question with a simple mathematical identity employing a very special variance of "Planck's constant." An estimation of the "gravitational constant" exists of course. Paul R. Heyl (1930) published the value of 6.670(5)×10–11 m3·kg–1·s–2 (relative uncertainty

0.1%),[28] improved to 6.673(3)×10–11 m3·kg–1·s–2 (relative uncertainty 0.045% = 450 ppm) in 1942.[29] A torsion balance was used to determine the approximated value above. A torsion balance is a particular type of tension spring that develops energy along its axis, as the ends are torqued inward. Same concept of the traps meant to dispatch rodents, vermin, etc.

The common value of "G force" was studied at great lengths, producing results of incredible accuracy. Published values of G derived from high-precision measurements since the 1950s have remained compatible with Heyl (1930), but within the relative uncertainty of about 0.1% (or 1,000 ppm) have varied rather broadly, and it is not entirely clear if the uncertainty has been reduced at all since the 1942 measurement. Some measurements published in the 1980s to 2000s were, in fact, mutually exclusive. Establishing a standard value for G with a standard uncertainty better than 0.1% has therefore remained rather speculative. By 1969 the value recommended by the National Institute of Standards and Technology (NIST) was cited with a standard uncertainty of 0.046% (460 ppm), lowered to 0.012% (120 ppm) by 1986. But the continued publication of conflicting measurements led NIST to considerably increase the standard uncertainty in the 1998 recommended value, by a factor of 12, to a standard uncertainty of 0.15%, larger than the one given by Heyl (1930).

The uncertainty was again lowered in 2002 and 2006 but once again raised, by a more conservative twenty percent, in 2010, matching the standard uncertainty of 120 ppm published in 1986. For the 2014 update, the Committee on Data of the International Science Council (CODATA) reduced the uncertainty to 46 ppm, less than half the 2010 value and one order of magnitude below the 1969 recommendation. [ref https://en.wikipedia.org/wiki/Gravitational_constant] This was/is humanity's means of space travel. Though no clear quantum quantification exists, this work poses exactly that. Though the approach outlined below is a much more tailor-made mathematical confection of sorts. A true quantifiable "identity" utilizing the proposed means of mediation. This is the only appropriate measure when identifying the empirical properties of any system. In 1589 Johannes Kepler began his university studies [ref 1 https://www.britannica.com/biography/Johannes-Kepler/Keplers-social-world] that would ultimately lead to a mentoring by the famous Michael Maestlin (1550–1631). This would inscribe a lifelong passion on young Kepler, the indelible mark of a true visionary.

Kepler would later go on, by use of Tycho Brahe's astronomical data (the best in existence at that time), to decipher the rudimentary planetary physics still used by NASA today to navigate the International Space Station (ISS) or negotiate shuttle travel. The work done by Kepler would lead to Sir Isaac Newton's great work. On July 5, 1687, Newton published his *Philosophiæ Naturalis Principia Mathematica* (in Latin). The *Principia* states Newton's laws of motion, forming the foundation of classical mechanics; Newton's law of universal gravitation; and a derivation of Kepler's laws of planetary motion (which Kepler first obtained empirically). [ref 2 scihi.org/isaac-newton-principial. Kepler's body of work was accomplished with brute-force computational rigor alongside primitive combinatorial mathematics. This incredible fact inspired my work, "Those who see furthest, suffer by candlelight."

# PART ONE

# GENERAL PLANETARY SPACING

---

## Outline

Exoplanet distribution has been a long-standing mystery, with many observations made and conclusions drawn. Though no clear universal understanding has been presented, this paper does exactly that, in the way of isolating a geometric numeric series that yields a near-exact reproduction of the innermost planets of our own system and several others. The most challenging portion of this collaboration was assembling a database of detailed star systems with a sufficient number of exoplanets to truly isolate the pattern of distribution. The data will show indisputable results and clear connections with respect to the spacing of the exoplanets in question and incidentally the orbital deviations of said exoplanets.

# Abstract

Planetary distribution has long since obeyed the Titius–Bode law   [ref   https://www.britannica.com/science/Bodes-law] (sometimes termed just Bode's law), which is a formulaic prediction of spacing between planets in any given solar system. The formula suggests that (extending outward) each planet should be approximately twice as far from the sun as the one before. The hypothesis correctly anticipated the orbits of Ceres (in the asteroid belt) and Uranus but failed as a predictor of Neptune's orbit. It is named after Johann Daniel Titius and Johann Elert Bode. Planetary formation occurs with the birth and subsequent life cycling of the epicenter of the system in question. The "epicenter" of course being the stellar center of each system. The sun and the planets formed together 4.6 billion years ago from a cloud of gas and dust called the solar nebula. A shockwave from a nearby supernova explosion probably initiated the collapse of the solar nebula. The sun formed in the center, and the planets formed in a thin disk orbiting around it. The rotational synchronization of each planet occurs from the phenomenon of centrifugal force, not unlike a synchronized dance.

Similarly moons formed, orbiting the gas giant planets. Comets condensed in the outer solar system, and many of them were thrown out to great distances by close gravitational encounters with the giant planets. After the sun ignited, a strong solar wind cleared the system of gas and dust. The asteroids represent the rocky debris that remained [ref https://

www.amnh.org/exhibitions/permanent/the-universe/ planets/formation-of-our-solar-system].

Ceres, however, is not a planet. Ceres is disqualified from planetary status because it does not dominate its orbit, sharing it as it does with the thousands of other asteroids in the asteroid belt and constituting only about forty percent of the belt's total mass.[44] Bodies that met the first proposed definition but not the second, such as Ceres, were instead classified as dwarf planets (https://solarsystem.nasa. gov/planets/dwarf-planets/ceres/in-depth/). A new computer simulation suggests that the gravity of gas giant Jupiter may have flung dwarf planet Ceres toward the sun during the volatile era of planet formation 4.5 billion years ago. This fact specifically precludes Ceres in the way its position is purely based on utter accident. This data would suggest that there is no remote connection between Ceres and the formation of our star system.

There has always been something out of place about Ceres. At 600 miles (1,000 kilometers) wide, Ceres is by far the largest body in the asteroid belt, the region between the orbits of Mars and Jupiter where space rocks (most of them only tens or hundreds of meters in size and oddly shaped) gather. Round like a planet, Ceres also contains some odd chemical compounds, such as ammonia, that are not present in its neighbors. Ceres's strange nature has long led scientists to believe that the dwarf planet is an intruder in the asteroid belt. A new simulation led by researchers from São Paulo State University in Brazil has now revealed a mechanism that may have displaced Ceres from its original birthplace in the distant past[ref https://www.space.com/simulation-reveals-birthplace-of-mystery-planet-ceres].

General debris is simply excluded from this integer sequence—parameters that adhere to the highest level of rigidity and produce results of the highest caliber. Planetary distribution is just that: the distribution of celestial bodies. Piecing the system together with objects of irrelevant categories defies and is a foul mistake on the road laid out by the scientific method. Thus Bode's law is invalidated on the clear understanding that Ceres does not obey the parameters of other celestial bodies; hence Ceres's existence is irrelevant. Ceres's position is not valid, thus not predicted in the sequence this paper focuses on.

The Titius–Bode approach ignores a much more deliberate sequence, which dictates specific ratios with respect to equally specific distances while imposing equally specific correlated orbital durations or each exoplanet distributed from its source. For any questions that would suggest that the series is incomplete because it does not reveal the deviation of the outer planets (Saturn, Uranus, and Neptune), the data would indicate that planetary systems of this size (as mentioned above) are entirely unique. Which would suggest that the series predicts that very fact. Simply put, orbital companions of the particular distance of the outer planets only exist due to entirely unique circumstances. The tautological fact of the series being finite, at exactly eleven terms of value, expresses clearly: this is a bound of formation, which applies to the number of celestial orbital bodies that naturally occur in star systems of variant mass. These particular values nearly perfectly outline our own system (as shown below), which would suggest this particular system to be of the highest stability. The habitation of Earth speaks volumes on that clear fact's behalf, coupled with the low level of eccentricity in each exoplanet's orbit, excluding Mercury—which is clearly an outlier with respect to its semi-major axis and orbital deviation.

The numeric series proposed here is anti-logarithmic. Meaning the spacing between each exoplanet does not follow an evenly spaced pattern, as the Titius–Bode law asserts. The reason for Mercury's absence from this explanation is simply because Mercury's semi-major axis and orbital deviation do not comply with the numeric series proposed here. In this way this numeric series acts as a template or blueprint of efficiency. Though as mentioned constructing a solar system that fits these particular deviations perfectly is rather unlikely. Orbital resonance, captured companions, and unstable orbits are simply a few variables that create variance among each star system. Though as will be demonstrated, the semi-major axis and orbital deviations proposed by this numeric series are completely universal, with little to no deviation, and spectacularly accurate. Just as in the explained exclusion of Saturn, Uranus, and Neptune, this work attempts to peel away articles of distraction and uncover the core means of planetary distribution in an entirely original way.

As mentioned below it is interesting to note that term number 10, namely 26/16, is the only term that can be reduced, and yields fibonacci integers. The reason arithmetically of course is because both the numerator and denominator of term 10 are even and have a common denominator (2). Reducing 26/16 yields 13/8. Both 13 and 8 are Fibonacci integers, as seen above. The clear message this numeric phenomenon details is simply the value of term 10 is the intended value for that particular quotient. Though the value does not theoretically converge until the nth term, 13 and 8 are both integers that yield the value (immature due to the integers' proximity to zero) according to the Fibonacci sequence. However the remaining terms, namely terms 1–11, excluding 10, also being approximations of the intended values; thus the integers in each quotient of the remaining terms are not "true" connecting integers to the intended values. That being said, even though the remaining terms are not "true" connecting integers, the values are quite clearly approximations to derivations of 1.62.

**Fig 1**

# Origin

| Deviation of semi-major axis | Deviation of period |
| --- | --- |
| 71/70 = 1.0142 | = 1.029 |
| 70/68 = 1.029 | = 1.046 |
| 68/65 = 1.046 | = 1.065 |
| 65/61 = 1.065 | = 1.089 |
| 61/56 = 1.089 | = 1.12 |
| 56/50 = 1.12 | = 1.16 |
| 50/43 = 1.16 | = 1.22 |
| 43/35 = 1.22 | = 1.34 |
| 35/26 = 1.34 | = 1.62 |
| 26/16 = 1.625 | = 2 |
| 16/5 = 3.2 | = 6.4 |

## Star System Border

*Integer sequence that dictates planetary spacing from the point of origin, (n/n-1), (n-1/n-2), (n-2/n-3), …, (n-7/n-8), (n-8/n-9), (n-9/n-10), (n-10/n-11).*

Interestingly the only terms in the series that can be simplified and are actual Fibonacci integers are 26/16 = 13/8 = 1.625 $\phi$.

There exists no simple quotient (unlike $\phi$) for 2/$\phi$, $\phi$^2/$\pi$, or 2$\phi$. Thus the terms yield values nearly exactly as the phi derivatives.

$$\underline{50/43} = 1.16 = 1.162 \Rightarrow \sqrt{} \; \phi^2/\pi$$
$$\underline{43/35} = 1.22 = 1.23606 \Rightarrow 2/\phi$$
$$\underline{35/26} = 1.34 = 1.358 \Rightarrow \phi^2/\pi$$
$$\underline{26/16} = 13/8 = 1.625 \Rightarrow \phi$$
$$\underline{16/5} = 3.2 = 3.236 \Rightarrow 2\phi$$

This particular numerical sequence is branded *iota*. The Greek letter iota is derived from the Phoenician letter *yodh*, meaning hand. Iota of course also has the meaning of small or infrequent. I believe these particular characteristics gallop parallel with the focus and meaning of the work it reflects. Meaning this particular orbital pattern described is difficult for star systems to demonstrate due to many extenuating variables that influence the compositional stage. As stated the data shows conclusively that our particular system displays this orbital resonance chain closer than any other known system. This fact is a fascinating prospect for potential habitation.

**Fig 2**

$$\iota = \left[ \frac{n}{n-x} \, , \; n = 71, \; \lim_{n \to 0} \int_{1}^{11} x \right]$$

*Mathematical notation expressing the limit of iota. As stated above n has an original value of 71 and subsequently loses value equal to the summation of x. As n approaches zero, x approaches 11. By nature of the integral, the function produces eleven terms.

This finite series dictates the proportionate spacing of planets (as well as orbital deviation) from their focal point. The denominator of the first term, 71/70, becomes the numerator of the successive term. Thus the denominator of the successive term is the numerator minus x (which is an ascending value from 1 to 11), and so on. In a similar style to the Fibonacci sequence, a "partition effect" is realized, and each term is "built" off the preceding terms. This is not to say every planetary system will have exoplanets displaying direct relationships in every instance in the series. The majority of planetary systems have one to three exoplanets (https://www.nasa.gov/image-feature/ames/planetary-systems-by-number-of-known-planets). Though this distribution pattern is entirely ubiquitous, coupled with the fact that only two systems known have the maximum number of exoplanets, our system and Kepler-90. The data clearly suggests that systems of that size are an anomaly.

**Fig 3**

*There exist many examples of this ubiquitous planetary spacing. These are a few (of many) that deviate in accordance with the suggested results. Solar mass/radius has no effect on results.

- Gliese 1002
- Gliese 357
- XO-2
- L 98-59
- HD 133131
- Kepler-138
- HD 5319
- Kepler-25
- V1298 Tauri
- Kepler-102
- Kepler-37

https://www.nasa.gov/

The results have a slight variation system to system, though the yields are always within the SD 10% ±.

*Example 1

---

## Our System

| | | |
|---|---|---|
| Venus – Earth | - | 1au/.723 au = 1.38.. |
| Orbital deviation | - | 365/224 days = 1.62222 |
| | | |
| Earth – Mars | - | 1.57 au/1 au = 1.57.. |
| Orbital deviation | - | 687/365 = 1.9 |
| | | |
| Mars – Jupiter | - | 5.2 au/1.57 au = 3.3 |
| Orbital deviation | - | 4333/687 = 6.5 |

---

*Example 2

---

## Gliese 1002

| | | |
|---|---|---|
| gliese b – c | - | .0457 au/.0738 au = 1.614... |
| Orbital deviation | - | 21/10.5 days = 2 |

---

## *Example 3

### Gliese 357

gliese b – c            -      .061 au/.035 au = 1.7..
Orbital deviation       -      9/4 days = 2.25

gliese c – d            -      .204 au/.061 au = 3.3..
Orbital deviation       -      55.5/9 days = 6.3...

## *Example 4

### XO-2

*XO-2 is a binary star. It consists of two componentsː XO-2S (also known as XO-2A) and XO-2N (also known as XO-2B).

XO b – c                -      .47 au/.14 au = 3.5..
Orbital deviation       -      120/18.7 days = 6.42222..

*Example 5

## HD 133131

*HD 133131 is a binary star in the constellation of Libra.

| | | |
|---|---|---|
| HD b – c | - | 4.79 au/1.4 au = 3.32.. |
| Orbital deviation | - | 3925/649 days = 6.1 |

*Example 6

## Kepler-138

*Kepler-138 is a red dwarf.

| | | |
|---|---|---|
| Kepler 138 b – c | - | .091 au/.075 au = 1.22.. |
| Orbital deviation | - | 13.7/10.3 days = 1.33.. |
| | | |
| Kepler 138 b – c | - | .1288 au/.913 au = 1.4.. |
| Orbital deviation | - | 23/9 = 13.7 days = 1.67... |

## *Example 7

# HD 5319

HD b – c       -     1.94 au/1.57 au = 1.235..

Orbital deviation    -     877/639 days = 1.37...

## *Example 8

# Kepler-25

Kepler 25 b – c     -     .11 au/.068 au = 1.617..

Orbital deviation    -     12.7/6.3 = 2...

## *Example 9

### V1298 Tauri

| V1298 b – c- | - | .108 au/.0825 au = 1.31... |
| Orbital deviation | - | 12.4/8.2 days = 1.52 |
| | | |
| V1298 c – d- | - | .1688 au/.1083 au = 1.57 |
| Orbital deviation | - | 24.3/12 days = 2.1... |

## *Example 10

### Kepler-102

| Kepler 102 b – c | - | .067 au/.055 au = 1.218.. |
| Orbital deviation | - | 7/5.2 days = 1.346.. |
| | | |
| Kepler 102 c – d | - | .086 au/.067 au = 1.29 |
| Orbital deviation | - | 10.3/7 days = 1.5... |
| | | |
| Kepler 102 d – e | - | .1162 au/.086 au = 1.35... |
| Orbital deviation | - | 16.2/10.3 days = 1.6... |

***Example 11**

## Kepler-37

Kepler 37 b – c     -     .139 au/.102 au = 1.36..
Orbital deviation     -     21.3/13.2 days = 1.605...

Kepler 37 c – d     -     .21 au/.139 au = 1.51...
Orbital deviation     -     39.3/21 days = 1.9...

https://www.nasa.gov/

*The magnitudes of the final semi-major axis and orbital deviation are 3.2 and 6.4, respectively. The concept is incredibly accurate and reproduces itself in countless star systems. This particular prediction gives enormous credence to the legitimacy of the proposed sequence.

**Fig 4**

## Explanation

The orbital deviation that results from the particular spacing yields a "partition effect" in the same way (with a subtle difference) as the Fibonacci sequence. Taking the sum of any successive terms divided by the third term in sequence yields rather uninteresting irrational integers.

$(1.16 + 1.23)/1.358 = 1.77$

$(1.23 + 1.358)/1.618 = 1.6$

$(1.358 + 1.618)/2 = 1.48$

$(1.618 + 2)/6.4 = .6$

*Though by taking the mean of each value yields:*

$(1.77 + 1.6 + 1.48) = (4.86/3) = 1.62$

$(1.77 + 1.6 + 1.48 + .6) = (5.46/4) = 1.36$

$(1.6 + 1.48 + .6) = (3.68/3) = 1.23$

*The same values are displayed.*
*This is a geometric connection that involves the operation of the first three terms, then all four terms. Then finally the three remaining terms.
*Also

$(1.162) \times (1.236) \times (1.358) \times (1.618) \Rightarrow 3.2$

$(1.162) \times (1.236) \times (1.358) \times (1.618) \times (2) \Rightarrow 6.4$

The products of the successive terms approach the final semi-major axis and orbital deviation, respectively.

## Summation

Numeric sequences govern the majority of what is complex, discreet, and by and large totally overlooked. The power and depth of an arithmetic/geometric pattern creating an echo into the physical universe is the most tremendous display of power man has come to know. Since the time of Euler, solving the great mysteries of the previous masters, humankind has employed the simplistic power of nature in architecture, transport, medicine, etc. This sequence accurately predicts planetary spacing from the point of origin (including orbital deviation), along with incidentally predicting the clear average range of star systems with respect to the number of planetary companions.

In the process of doing so, a new Fibonacci-like sequence has been revealed. One that shows not only the connection between phi and planetary spacing, but also other derivatives of phi. The main terms in the sequence include $(\sqrt{\phi^2/\pi})$, $(2/\phi)$, $(\phi^2/\pi)$, $(\phi)$, and $(2\phi)$. Interestingly the only terms in the series that can be simplified and are actual Fibonacci integers are $26/16 = 13/8 = 1.625 \Rightarrow \phi$. There exists no simple quotient (unlike $\phi$) for $(\sqrt{\phi^2/\pi})$, $(2/\phi)$, $(\phi^2/\pi)$, or $(2\phi)$. Thus the terms yield values nearly exactly as the phi derivatives. The deep connection displayed here shows overwhelming clarity with respect to the harmonic splendor of nature. The precision of the results is astounding, with little to no deviation from the proposed results system to system. Particular aspects of this sequence at first glance were incredibly bewildering. Though as the layers were slowly removed, a clear pattern formed. The sequence itself, which describes the semi-major axis / orbital deviation, was discovered simply through combinatorics.

## References

https://www.britannica.com/science/Bodes-law

https://www.amnh.org/exhibitions/permanent/
    the-universe/planets/formation-of-our-solar-system

https://www.britannica.com/biography/Johannes-Kepler/
    Keplers-social-world

https://solarsystem.nasa.gov/planets/dwarf-planets/ceres/
    in-depth

https://www.space.com/
    simulation-reveals-birthplace-of-mystery-planet-ceres

https://www.nasa.gov/image-feature/ames/
    planetary-systems-by-number-of-known-planets).

# PART TWO

# HABITABLE EXO-PLANET NUMERIC IDENTIFICATION

## Outline

Habitable exoplanet zones are cosmological equilibrium points. A point in which liquid H2O or water can maintain its state with respect to the eccentricity of the planet's orbit, solar luminosity, etc. Nearly 4,000 planets have been categorized as "habitable." These points are colloquially referred to as "Goldilocks points."

Obviously radiant energy output is proportionate to the overall stellar mass of whichever system is being studied. Not all star systems are created equally for bearing habitable exoplanets; K dwarfs are the true "Goldilocks stars," said Edward Guinan of Villanova University, Villanova, Pennsylvania. "K-dwarf stars are in the 'sweet spot,' with properties intermediate between the rarer, more luminous, but shorter-lived solar-type stars (G stars) and the more numerous red dwarf stars (M stars). The K stars, especially the warmer ones, have the best of all worlds. If you are looking for planets with habitability, the abundance of K stars pump up your chances of finding life." [ref https://exoplanets.nasa.gov/search-for-life/habitable-zone/]. These "circumstellar habitable zones" are expressions of incredibly improbable variables flowing in near-perfect concert for billions of years without falter. Taking into consideration the magnitude of the potential margin of error, one reflects on very clearly the reason only one planet (our own) has spawned habitation. There are many avenues of metaphysical dissent that royally would claim zones such as these the proof of intelligent design. But with the facts in considerable availability, the circumstances are simply ones of remarkable complexity and chance.

And remarkably an integer that defines life and habitation at the quantum level, simultaneously involving itself in the affairs of the heavens. This pervasive set of characteristics lends insight into the efficiency of nature. A very metaphysical concept, efficiency. A representation of the most logical method of performing a certain task A. Our certain task A dictates itself to be the most logical path of travel: A = A, tautological. Thus the Fibonacci integer A is the shortest distance to its goal B, thusly A to B = A. The results shown in this portion display an unprecedented and deeply esoteric numeric relationship. The idea of the golden ratio acting in concert with the solar radius and mass within specific parameters of conversion sheds light on the blurred border that separates the animate from the inanimate. The harmonious bonding of the logical and the sentient. The original prospect for the Fibonacci integer series was focused on the breeding habits and cycling of hares. This particular pattern was etymologically isolated, deconstructed, and explained in terms of real-world instances. This work is built on that same philosophy, building the truth from facts. The mortar is simply time.

## Abstract

The golden ratio, deemed the most ubiquitous integer, expresses the highest level of efficiency within a natural mechanism. Nearly a decade ago, a team of researchers discovered an incredibly complex system of deviation with respect to lengthy intervals of time and the number of revolutions each planet within a particular system exhibited [*ref https:// stillnessinthestorm.com/2015/05/all-solar-system-periods-fit-fibonacci/*]. The results yielded the frequency of revolutions (among other variables) of the celestial bodies correlated to the Fibonacci sequence. This requires an enormous amount of computation and data. Our particular system was chosen for this study based on the availability of raw information, coupled with the very details (with respect to the unique number of celestial bodies) our system offers. A more specific concentration of variables animates a clear connection between the golden ratio and habitable exoplanets. Isolating Fibonacci integers within planetary dynamics and the sun's unique solar radius will be our instrument of collection. Employing the radius of our own sun as a cosmic measuring stick of sorts allows a hidden numeric phenomenon to become visible. By converting the semi-major axis (as explained below) of the exoplanet in question, one can see a clear pattern, specifically the semi-major axis of said exoplanet will be an incredibly close representation (sometimes the exact integer) of one Fibonacci integer. Then the orbital duration in Earth days serves as the remaining Fibonacci integer (a, b).

This method is simply focused on the semi-major axis of a particular exoplanet, which orbits its host at a specific radius. The "golden radius" is a distance expressed in *solar radius* units but is first converted to astronomical units (92,955,807 miles) proportionate to the stellar mass of the system in question. Calculating this radius is simple: mass units (ex. <u>Sol</u> = 1 mass unit, <u>Trappist</u> = .089 mass units) expressed *in astronomical units* and finally converted to solar radius units (432,700 miles per). This concert of coefficients—solar mass units, solar radius units—coupled at a specific semi-major axis yields Fibonacci integers (a, b). This is expressing the efficient relationship between these values with respect to the possibility of habitation. This is occurring simply because the habitable zones of each star system are proportionate with respect to solar mass, solar radius, and semi-major axis of a particular exoplanet. Assuming these parameters maintain continuity, the chance of habitation is possible. If a particular coefficient diverges, the habitable zone diverges as well.

Not all exoplanets in habitable zones exhibit the golden ratio. Though almost all exoplanets that exhibit the ratio are within the habitable zones. And the few exoplanets that yield Fibonacci integers (a, b) and do not lie within the habitable zone express the primitive habitable zone when the star was younger. Thus the habitable zone has diverged with respect to solar radius/mass. The quotient measuring the ratio of orbital period in Earth days and semi-major axis expressed in solar radius units, respectively, yields yet again the golden ratio. As shown below our particular planet of habitation (Earth) displays this very ratio. Meaning logically all exoplanets that express specifically Fibonacci integers (with respect of the orbital duration and the semi-major axis) reside at proportionate Earth-like semi-major axis, unique to the radiation output, luminosity, etc. That is of course proportionate to the stellar epicenter of the system in question. For exoplanets that reside in the habitable range and do not exhibit Fibonacci integers with respect to their orbital dynamics, they still display a specific quotient value. There exists a range of values that directly correlate to the ratio described above. Any value in the range of 1 and 1.618 (golden ratio) with respect to the orbital period in Earth days and the semi-major axis in solar radius units describes an orbital distance within that habitability zone. Unless the solar mass/radius ratio irreparably diverges as shown below.

**Fig 5**

_Randomly chosen potentially habitable exoplanets._

### K2-72 e

(A) Orbital period in Earth days = 24.15 days

(B) Semi-major axis = .160, AU = 22.77 solar radius units

A/B = 1.09. Clearly within the proposed ratio of habitability.

### GJ 1002 c

(A) Orbital period in Earth days = 21.2 days

(B) Semi-major axis = .0738, AU = 15.85 solar radius units

A/B = 1.337. Clearly within the proposed ratio of habitability.

### GJ 1061 d

(A) Orbital period in Earth days = 13 days

(B) Semi-major axis = .054, AU = 11.6 solar radius units

A/B = 1.12. Clearly within the proposed ratio of habitability.

*There of course exists a small deviation with respect to the exoplanets on the very inner "edge" of the habitability zone. The value of the proposed ratio can sometimes fall below the inner bound of 1. Though the deviation is negligible.

## Example
### *LP 890-9 c*

(A) Orbital period in Earth days = 8.45 days

(B) Semi-major axis = .0398, AU = 8.55 solar radius units

A/B = .98. Clearly within the proposed ratio of habitability.

## Fig 6

$$F_n = \left\{ \frac{M_{au}}{R} , P_d \right\}, \left\{ a = \frac{M_{au}}{R} , b = P_d \right\}$$

Calculating this semi-major axis of habitability is simple: solar mass units, expressed *in astronomical units (which is then converted to miles)*, converted to solar radius units (432,700 miles per). Then simply the orbtial period in days is the successive Fibonacci integer, with respect to the chronological Fibonacci numeric order.

There is a slight deviation in star systems <.7 solar masses and star systems >2.5 solar masses.

The solar mass to astronomical unit conversion has a slight deviation at times, though the results are still Fibonacci integers.

Fn, the Fibonacci sequence (a, b), is expressed at this particular radius in the way of: the semi-major axis of the exoplanet in question expressed in solar radius units (432,700 miles (a)) and the orbital period of the exoplanet in days (b).

Fn = {0, 1, 1, 2, 3, 5, 8, 13, 21, 34, 55, 89, 144, 233, 377, 610, 987, …, fn}

**Fig** 7

| Star system | Solar Mass | | Semi-Major axis | Orbital Period | a\|b |
|---|---|---|---|---|---|
| Gliese 514 (b) | .5 ˝ | .42 au | 89 Solar radius units | 142 d | ˝ 89,144 |
| Kepler 174(d) | .7 ˝ | .67 au | 144 Solar radius units | 246 d | ˝ 144,233 |
| Kepler 160 (e) | 1.1 ˝ | 1.089 au | 233 Solar radius units | 378 d | ˝ 233,377 |
| HD 112640(b) | 1.8 ˝ | 1.75 au | 377 Solar radius units | 613 d | ˝ 377,610 |
| HIP 65891(b) | 2.5 ˝ | 2.8 au | 605 Solar radius units | 1000 d | ˝ 610,987 |
| SD ± 15% | | | | | |

ref. https://exoplanetarchive.ipac.caltech.edu

Gliese 514 b, Kepler-174 d, and Kepler-160 e are all exoplanets within the habitable zone of their systems.

*Earth (217,365 ≈ 233,377)*
*Venus (150,223 ≈ 144,233)*
*Venus is also located in the habitable zone. Data indicates that Venus most likely had an ocean when the sun was younger and dimmer. Roughly 750 million years ago. Venus may have had a shallow liquid-water ocean and habitable surface temperatures for up to 2 billion years of its early history, according to computer modeling of the planet's ancient climate by scientists at NASA's Goddard Institute for Space Studies (GISS) in New York.

www.nasa.gov/feature/goddard/2016/nasa-climate-modeling-suggests-venus-m…

HD 112640 b and HIP 65891 are both primordial habitable exoplanets that no longer have the potential for life due to the massive divergence in solar radius/mass ratio.

**Fig 8**

$$\frac{R}{M}\,au$$

*A simplistic way of approximating the habitable zone divergence: factoring solar radius units with respect to solar mass units yields the approximated habitable "inner edge," sparking the nearly infinitesimal chance of life.

Solar radius/mass ≈ habitable zone divergence, expressed in AU (astronomical units)

- If the ratio solar radius/mass ratio is not ≈ 1, the habitation zone shifts accordingly.
- The solar radius is known to diverge from the solar mass with larger stars. The magnitude of radius–mass divergence dictates the divergence of the habitable zone.

**Examples**
**_Star HD 112640_**

(380,613 ≈ 377,610)—Fibonacci integer yield
Solar radius ≈ 39

Solar mass ≈ 1.8

39/1.8 ≈ 20 AU, estimated habitable zone

This is correct.

*ref. https://exoplanetarchive.ipac.caltech.edu*

### *Star Kepler-432*

(233,406 ≈ 233,377)—Fibonacci integer yield

Solar radius ≈ 4

Solar mass ≈ 1.3

4/1.3 ≈ 3 AU, estimated habitable zone

This is correct.

*ref. https://exoplanetarchive.ipac.caltech.edu*

### *Star HD 47536*

Solar radius ≈ 23.5

Solar mass ≈ 1

23.5/1 ≈ 23 AU, estimated habitable zone

This is correct.

*ref. https://exoplanetarchive.ipac.caltech.edu*

### *Star HD 177830*

Solar radius ≈ 3

Solar mass ≈ 1.5

3/1.5 ≈ 2 AU, estimated habitable zone

This is correct.

*ref. https://exoplanetarchive.ipac.caltech.edu*

### *Star HD 27442*

Solar radius ≈ 6.6

Solar mass ≈ 1.2

6.6/1.2 ≈ 5.5 AU, estimated habitable zone

This is correct.

*ref. https://exoplanetarchive.ipac.caltech.edu*


### *Star HD 222404*

Solar radius ≈ 4

Solar mass ≈ 1.5

4/1.5 ≈ 2.7 AU, estimated habitable zone

This is correct.

*ref. https://exoplanetarchive.ipac.caltech.edu*


### *Star HD 190984*

Solar radius ≈ 1.5

Solar mass ≈ .9

1.5/.9 ≈ 1.6 AU, estimated habitable zone

This is correct.

*ref. https://exoplanetarchive.ipac.caltech.edu*


All star systems diverged with respect to the ratio of solar radius to mass as stated. When this occurs the circumstellar habitability zone diverges accordingly. This simple quotient, performed with the specific conversions demonstrated, yields powerful results. The radius–mass quotient is a long-overlooked means of approximating radiant energy divergence points. The method even accounts for luminosity divergence, apparent gravitational wobble (with respect to binary systems), etc. The overall production is completely unfazed, even considering the onslaught of variable parameter augmentations. As shown above the solar mass units correlate nearly perfectly with the semi-major axis conversion. The semi-major axes of these select exoplanets in solar radius units (432,700 miles) produce a Fibonacci integer (a) corresponding to the satellite star's solar mass. While simultaneously the exoplanet's orbital period in days produces the successive Fibonacci integer (b). There are many other exoplanets that produce clear results, varying in solar mass. These subjects were chosen due to their extreme convergence to the Fibonacci sequence. Though by the nature of supermassive stars and their physical inability to harbor orbital companions, it seems there is a clear bound with respect to the value of the Fibonacci integers produced. In the way of stars with solar masses >2.5 (with planetary systems) are very rare; as shown in *Fig 7* expresses the proposed "range" of production.

This proposed "stellar bound" is very logical, in that there exists a clear pattern within the multitude of planetary systems with respect to the unique stellar mass of each. This particular range that exists, approximately .1 solar mass to approximately 2.5 solar masses as mentioned above, is an obvious harmonious scale of relative divergence with respect to nebulae production. Stellar epicenters exceeding 2.5 solar masses are simply too unpredictable with respect to their solar evolution to allow for stable, lengthy intervals of orbital alliance.

## Explanation

The golden ratio expresses the natural efficiency of a mechanism. As stated the particular stellar dynamics: solar radius and mass proportionately allow the habitable zone to occur at a proportionate semi-major axis. A divergence in solar radius/mass shifts the habitable zone, thus the emergence of the golden ratio expresses the efficient relationship between solar radius and mass and the semi-major axis of the exoplanet. All exoplanets which exhibit Fibonacci integers are positioned within the habitable zone. Interestingly in the particular instance of habitability, almost all exoplanets sharing these particular characteristics are positioned at the same proportionate semi-major axis as our Earth. Exoplanets exhibiting the Fibonacci integers that are not within the habitable zone are primitive habitable exoplanets. The solar radius/mass divergence destroyed the habitability of the exoplanet in question, the same fate our zone will face in time. These exoplanets act as cosmological time stamps, tracking the evolution of the star in question with respect to solar radius/mass and the margin of the habitable zone. In the instances of great stellar expansion, large portions of space are immersed in cosmic rays, leaving no remote trace of the original "circumstellar habitable zone" remaining.

Thus the same concept of altering a well-functioning mechanism, be it mechanical, organic, or silicon based in nature. There always exists emergent probabilistic properties that allow for unforeseen and radical means of dissent. The delicate balance of solar gravitational exertion and orbital exoplanet resonance is a phenomenon of unbelievable complexity. Existing in balance for sometimes billions of years before collapse, minor changes in a particular dynamic variable will have an enormously profound effect on the totality of any star system, given a long enough interval of time.

## Summation

This entirely unique numeric method allows for the simple identification of potentially habitable exoplanets while displaying a deep and evidently overwhelmingly pervasive relationship between natural mechanisms and the golden ratio. Even with the concept of simplicity and logical data flow, one still seems lost in translation. Universally there exists a "golden radius." The golden radius is calculated simply by taking the solar mass units of a particular exoplanet, converting the value to astronomical units, then to solar radius (432,700 miles) units. This variable concert is not dissimilar to the dispersal of angiosperm offspring or water droplet patterns. Ubiquitous and overwhelmingly efficient. Truly the codons of the physical realm, animate and inanimate. This function produces a Fibonacci integer (a), then simply the orbital period in days of the exoplanet at the specific radius produces the successive Fibonacci integer (b). This pair (a, b) is the proposed marker of habitability. And with the logical fact of proportionate stellar dynamics, this phenomenon partitions itself through each category of the stellar epicenter.

Just as the Fibonacci integers are present within collective orbital patterns, sunflower faces, leaf displacement on branching, etc., the Fibonacci integers also emerge concerning the concert of variables necessary for exoplanets to be potentially habitable initially. Put simply the spacing of the habitable zones in each star system is with respect to the golden ratio as well. Efficiency breeds efficiency, tautology. The integer itself expresses efficiency within natural mechanisms, from the micro to macro levels. But also expresses the harmonious relationships necessary to support life.

As mentioned every exoplanet that exhibits the particular Fibonacci characteristics with respect to semi-major axis in solar radius units and orbital duration in Earth days will fall within a proportionate Earth-like distance within the unique star system's habitability zone. With the exception of those exoplanets whose stellar epicenters have had considerable divergence in solar mass/radius. Though there were those exoplanets mentioned that fell within the proposed ratio of habitable distances of Fig 5 and exhibited a planetary dynamic ratio of >1 to approx. 1.625. This esoteric relationship is a powerful numeric correlation that runs parallel to the proposed habitability zone of each star system. This "mathematical translation" is projecting a tremendous message, that message being even the most complex concepts, including but not limited to the habitability of exoplanets, can be explained in the language of mathematics and logic.

## References

https://exoplanetarchive.ipac.caltech.edu
https://stillnessinthestorm.com/2015/05/
    all-solar-system-periods-fit-fibonacci/
www.nasa.gov/feature/goddard/2016/
    nasa-climate-modeling-suggests-venus-m…
https://exoplanets.nasa.gov/search-for-life/habitable-zone/].

# PART
# THREE

# CONCEPT OF GRAVITATIONAL ORIGIN/ PROPAGATION, ATOMIC DESTABILIZATION

---

(with Supplemental Explanation on the Existence of Dark Energy, etc.)

## Perpetual Energy

Perpetual energy may refer to perpetual motion, the property of an *imaginary device* that, once started, continues to move forever with no input of external energy.

# Introduction

Since the time of Newton, this question remains, "By what means does gravitation exhibit force between celestial manifolds?" Newton himself, after spending a lifetime constructing the mathematical tools for interpreting gravitational force, could not explain gravity in a simple and coherent manner. The means of the gravitational mechanism eluded Newton, as it has all humankind. Albert Einstein arrived in the world with the sole purpose of shattering the "typical" view of gravitation and overall the motions of all heavenly bodies. The mission for Einstein was clear: abridge the constructs of Newtonian physics with modern-day (Einstein's modern day of course) rigorous physics and produce clear, consistent results that could be then easily accepted by his colleagues.

After the "Miracle Year," in which Einstein produced four papers, he presented his famous paper on general and special relativity. The main concept being spacetime (a construct invented by Einstein to explain the distribution of spatial geometry) is an ever-present inter-woven physical construct, universal and without bound, which is deformed in the presence of mass. Specifically the deformation or depression of spacetime was most noticeable when a star celestial orbiter was present. Einstein explained that the deformation or curvature of spacetime would also affect waves of light photons. Einstein predicted that the light vector crossing through the gravitational "field" of a celestial body should be elongated or "bent" by gravitation. Sir Arthur

Eddington led an expedition to photograph the 1919 total eclipse of the sun. The photographs revealed stars whose light had passed near to the sun. Their positions showed that the light had been bent exactly as Einstein had predicted [ref https://www.physlink.com/Education/Askexperts/ae198. cfm#:~:text=Einstein%20predicted%20that%20light%20 should%20be%20bent%20by,had%20been%20bent%20 exactly%20as%20Einstein%20had%20predicted.]

Though this experiment does not technically constitute proof, the images captured gave a clear impression of light vectors curving around the circumference of the moon. Although this does not mean spacetime deformation was responsible for the path of those particular photon array aberrations. To quote Nikola Tesla, "Einstein's relativity work is a magnificent mathematical garb which fascinates, dazzles and makes people blind to the underlying errors. The theory is like a beggar clothed in purple whom ignorant people take for a king… its exponents are brilliant men but they are metaphysicists rather than scientists." The concept of general relativity is fundamentally conspired to elude the notion of dissent, setting the parameters of the spacetime construct to a level of angelic harmony. A blueprint of such divinity, one must not question even the parchment it's scorched upon. Taking all basic responsibility off the proverbial shoulders of the theory itself. How can a massless construct present resistance in the presence of mass when the very laws of cause and effect make any action in this particular scenario immutable to dissection?

Spacetime is no more illusory than religion itself; the entirety of the philosophy is built on faith alone when the overwhelming data has oriented science in pursuit of explanations at the quantum level. That is the sole mission of the final portion: to unearth the clear identity of gravitation and incidentally expose the relationship, in which gravitation and dark matter/energy share.

## Outline

This ridiculous anti-Newtonian "concept" (_perpetual energy_) is the key characteristic that will aid in eventual proof of an emphatic dissent with respect to the continuous origin of gravitational fields. The core understanding of Einstein's imagination of gravity includes the "deformation" of spacetime in the presence of mass. As quantum physics discovered and presented, more than ninety-nine percent of all atomic energy—the force responsible for maintaining an atomic structure's integrity—is allocated to "binding energy." Binding mass is simply the continuous frequency that holds the "form" of the particular quantum arrangement in question. All elements forged in the heart of stars are subject to the extreme pressure of fission, the amount of energy required to separate a particle from a system of particles or to disperse all the particles of the system. Binding energy is especially applicable to subatomic particles in atomic nuclei, to electrons bound to nuclei in atoms, and to atoms and ions bound together in crystals.[ref https://www.britannica.com/science/binding-energy].

Nuclear binding energy is the energy required to separate an atomic nucleus completely into its constituent protons and neutrons or, equivalently, the energy that would be liberated by combining individual protons and neutrons into a single nucleus. The hydrogen-2 nucleus, for example, composed of one proton and one neutron, can be separated completely by supplying 2.23 million electron volts (MeV) of energy. Conversely when a slowly moving neutron and proton combine to form a hydrogen-2 nucleus, 2.23 MeV are liberated in the form of gamma radiation. The total mass of the bound particles is less than the sum of the masses of the separate particles by an amount equivalent (as expressed in Einstein's mass–energy equation) to the binding energy.

With this fact in mind, the idea of gravitation as a _mechanism_ (a natural or established process by work is done and production is realized) sheds an interesting light on the invariable and immutable laws of reality. The construct of the gravitational field relies on "likeness" attraction. Quite the opposite expressed on the atomic level, considering similar subatomic particles are less than "receptive" to each other. Protons naturally bonding with electrons, and vice versa. Examining the concept of relativistic gravitation, one is left imagining a spherical body of considerable mass. The understanding being: within a particular volume proportional to the mass of the body in question, there exists a "field" that is animated simply by the presence of the mass itself. Realistically this is envisioned as the "curvature of spacetime," topological deformation of what is normally a "straight" vector. The more massive the celestial object in question is, the greater the magnitude of the deformation to the vector in question.

Problems with the relativistic gravitational approach:

- The curvature of spacetime addresses the cause, i.e., the elongation and thus curvature of a vector ($v$), though not the mechanism of effect (aside from the deformation of empty space, concept).

- In accordance with problem one, no mention of energy loss is considered with the invitation of relativity. An interaction must be occurring within the realm of the physical, meaning simply an even exchange of energy should be present. This categorically excludes the concept of spacetime on the simple fact that spacetime has no rest mass. The proposed variant mass described in the reaction of depression would violate the mass/energy conservation construct.

When positive work is done on an object, the system doing the work loses energy. In fact the energy lost by a system is exactly equal to the work done by the system. Like work (W) the unit of energy (E) is joule (J). With respect to energy loss, a particularly unique situation could exist in which the force driving the machine could preserve the energy being applied with minimal energy loss over time.

Which is precisely the issue posed in this outline. The concept relies on the mass of the body in question to cause physical deformation by means of no energy transfer other than mere existence. This approach hurdles over many questions and larger inadequacies. The major hurdle facing relativity at the moment is reconciliation with the quantum constructs. Identifying the mediation of energy exchange with respect to gravitation will come from the examination of forces such as electromagnetism, strong/weak force, etc., to pose a common ancestry.

A common ground must first be achieved in the way of complete acceptance of the empirical truth that all matter is essentially motion. The motion begins at the quantum level and resonates collectively, creating a frequency of solid state. Be it matter or antimatter, this must be a universal conception. Having established a universal truth, "Motion is the universal engine of matter," we navigate further. Assume different variations of subatomic mechanisms exist, which they absolutely do. We have established "motion" to be the engine of both mechanisms; now we must establish the universal mediation in the universe. Simply put, if photon arrays are the universal currency with respect to electromagnetism and the photoelectric effect, etc., then what would be the universal means of currency exchange? That is simply the goal of this work: to fill the void left by a generation of great men, with rather narrow visions at times, who refused logical concessions while being presented with conclusive data. Posing a more realistic approach with harmonious fettering is an art form; one must perfect the negotiations over the peaks and troughs of exploration.

## Considering the Quantum

It is known that gravitational waves (which have been detected) move at light speed, 186,000 miles/second. Special relativity emphatically expresses how a photon stands alone with respect to velocity/speed. I'm paraphrasing of course, but the general idea resides amid the comic relief. This logically proves photons or quantum particles are the mediating fuel that drives the gravitational engine. But the question now arises, "By what means is gravitation developed?" This is a question again best answered by logic. Imagine the concept of the "Big Bang." This is the most reliable simulation, so it's the best model for this particular thought experiment. At the moment the primordial mass of the universe began to expand and cool, the gravitation would have been obviously so great, no progress would have been made. And a static molten pulp would exist today.

Interestingly quite the opposite occurred. The hadron epoch (period of particle assimilation) details that seconds after the supposed "Big Bang," protons began forming, followed by electrons, neutrons, etc. This subatomic mingling would suggest little to no constraint on the flow and divergence of newly forged particles. This clearly suggests the particular circumstances that allow gravitation presently did not exist at the moment of subatomic birth. There are now only two logical solutions to this chicken-or-egg-like quandary:

- Particular subatomic particles (undetected) are directly responsible for the phenomenon of gravitation by means of field construction or energy transfer.

- Dark matter/energy cooled much later than visible everyday matter due to atomic structural differences. Theoretically dark matter still could be cooling and evolving atomically to this very day, implying the relationship between matter and antimatter is directly responsible for gravitation due to the effects of gravitation seeming to manifest in response to dark matter's emergence.

There exists of course the scenario (however improbable) of curvature or deformation of spacetime being the actual explanation of the force concerning gravitation. This explanation, as demonstrated (deformation of spacetime), encounters singularities (points of functional disjointment) in circumstances in which enormous amounts of mass reside within a single manifold. For example the "Big Bang," the center of supermassive black holes, etc. These known phenomena must adhere to a rigid scientific framework that encounters no remote deviation (regardless of the insignificance of impact). These clear inequities in relativity can be amended with the induction of quantum theory.

Electromagnetism is the essential action of mediation in magnetism. Photon arrays, ultraviolet lights, etc., act as carriers or mediators of energy for particular subatomic structures. The effect being described is the same effect that allows for the Earth's electromagnetic field to shield the planet from harmful cosmic rays. Electricity and magnetism are two aspects of electromagnetism as mentioned above. In 1905 Einstein's special theory of relativity established beyond a doubt that both are aspects of one common phenomenon. At a practical level, however, electric and magnetic forces behave quite differently and are described by different equations. Electric forces are produced by electric charges either at rest or in motion. Magnetic forces, on the other hand, are produced only by moving charges and act solely on charges in motion. [ref https://www.britannica.com/science/electromagnetism].

So the concept of both actions (electricity and magnetism) operating in concert creates an unimaginable amount of raw energy (photon arrays) to flow in the path of attraction. The net force of every atomic mechanism in the structure emitting a flow of an energy field stimulates the attraction of oppositely charged particles. The key concept here is the indisputable fact that light or photon arrays mediate the force of electromagnetism. Thus in order to isolate the allocation of potential gravitational per atomic mechanism, one must employ the appropriate constant to mediate the value just as accurately as energy transfer is navigated. Planck's constant describes the relationship concerning a single photon's energy to its frequency of peak and trough. Planck's constant was the solution to the Rayleigh–Jeans catastrophe. Essentially a particular bound of radiation emission was proposed by the aforementioned researchers. The value proposed was overestimated, as the world of science was about to realize. At particularly low frequencies, the proposed value would heavily diverge.

Max Planck derived the constant harmonizing the frequency and energy emission of a light wave: 6.62607015×10–34 J·Hz. Angular momentum is the driving force of photon arrays. The cyclical pattern of travel is both efficient with respect to energy conservation and necessary for logical means of motion. A particular conversion of Planck's constant yields the "reduced Planck's constant," which is simply the quotient of Planck's constant with respect to $2\pi$. This quantification of Planck's constant allows for measurements of the magnitude of energy, in particular "lumps" of angular momentum within the vector of the photon array in question.

The equation below employs this integral quantification by means of two times the reduced Planck's constant minus 1. And then simply the thousandth of that particular value is extracted. This value is then finally multiplied by the average atomic mass of the celestial body in question and the number of atomic mechanisms present in the manifold. Unlike the "gravitational constant," which is a product of the mass of the body in kilograms and the constant itself, this identity isolates the estimated allocation of energy per atomic mechanism by utilizing only the one constant by means of simplistic arithmetic operation. The concept of "mix and match" unrelated constants to achieve an estimated value is disingenuous. There exist innumerable combinations to achieve any particular integer, though finding a constant that is directly animated by the concept itself and then employing a simplistic and elegant operation to achieve a value less than one-thousandth the value predicted by the "gravitational constant."

**Fig 9**

$$\left[ \frac{\left(2\frac{\hbar}{2\pi} - 1\right)}{10^3} u_x \Big| a_x \right] \approx G$$

*Planck's constant is expressed in Joules. Joules interestingly exist within a 1:1 ratio with newtons! A seamless conversion, adding validity to this proposed identity.

As mentioned the identity above, expressed in newtons (kilogram/meter/second), represents the gravitational energy emitted from each atom due to repellant photon arrays allocated to all atomic mechanisms within whichever celestial manifold is a coefficient of particular interest; hence (ax), meaning the number of atoms in x. Because the identity is a coefficient, the energy output for each subject is variable.

The concept of atomic attraction through electromagnetism is described by "Coulomb's law." This law dictates the laws of attraction between charged particles. The notations of both Coulomb's law and Newton's law of gravitation are identical, which has been proposed in the past as showing a clear link between quantum and gravitation. The relevance of this is specifically considering the force exhibited on each atomic mechanism, from a 360-degree array or more specially, and anti-divergence. Instead of a conical dispersal phenomenon, seen in light waves, gravitation, radio waves, etc., a concentrated emission from every available differentiated outward point "wraps" around each individual subatomic particle like a python strangle.

This is the point of our original quandary concerning the engine propagating gravitation. When considering a specific subatomic particle as the catalyst of animation, we consider the magnitude of matter necessary to exist that would create the sizable emission of radiation that explains a force as ubiquitous as gravitation. This avenue of thought leaves our conclusion posted on a single doorstep, that of dark matter. Dark matter/energy encompasses nearly ninety percent of the matter in the universe while seemingly having little to no physical reactivity. This work focuses on the plausibility of dark matter acting as a cradle or means of cosmic transport for the visible matter.

Simply imagine dark matter as the vehicle in which matter is stored, controlled, and transported through the cosmos while ensuring the stability of the cargo in question.

Make no mistake though, the means of constriction via dark matter and visible matter are anything but placid. The basics of electromagnetic attraction explain that instances of photon arrays traveling in favorable directions with neighboring energy cloud emissions allow for the force of subatomic attraction. On the contrary the details surrounding attraction nullify themselves through mutual expulsion of differential energy dispersal. Though imagining dark matter in the way of being compositionally oppositional to all matter or variant atomic makeup, meaning the photon arrays of dark matter atomic structure would always be moving in the opposite direction as all energy dispersals. The plausibility of this means of operation seems low when first considered. Though deep connections begin tethering in this supposed realm. In a mathematical respect, dark matter can be seen as a complex number. A number with a "real" portion and an "imaginary" portion.

As shown in *Fig 10* below, the concept of one atomic mechanism that is the makeup of dark matter would be in two simultaneous states of emission. A Euclidean matrix (expressing the integrity of the dark matter mass) would depict the state of repulsion the particle would be in while locked in a vacuum. A theoretical state with no other mass interpolating. Now imagining the introduction of dark matter or otherwise, the state of emission would extend to the "complex plane" or more simply deform the Euclidean plane in which the manifold resides. This will sound familiar, the concept of deformation of a universally present body.

Though the obvious augmentations are evident, data and clear logic will allow the connective shreds of cosmic truth to coagulate. This concept of "universal repulsion" raises questions of considerable relevance: "Is dark matter attracted to itself?" and "Is dark matter repelled by photon arrays?" Both are very important questions that must be addressed. Answering respectively dark matter is in all probability repelled even by other dark matter/energy. This would explain the rapid acceleration concerning the visible universe. The Hubble telescope project was originally launched in 1990 to gather images of deep space. The telescope was named after Edwin Hubble.

Edwin Hubble originally made the discovery of light frequency variations at great distances. This phenomenon, denoted "red shift," gives an approximation of the age of a particular light vector based on spectrum emission. Though after the launch and data collection of the Hubble telescope, a shocking realization was revealed. Riess and collaborators received the Nobel in 2011 after Hubble and other observatories confirmed that the universe was accelerating in its expansion. Riess calls this latest Hubble effort a "magnum opus," given that it draws upon practically the telescope's entire history, thirty-two years of space work, to deliver an answer. [ref https://www.space.com/universe-expansion-rate-hubble-telescope-measurements].

The second, and more challenging, question to address is, "Is dark matter repelled by light arrays?" The obvious concepts gained by the intuitive naming of "dark matter" would imply the matter itself is rather camera-shy. That being said, light or photon arrays are universal ingredients of mass. E=Mc^2, the energy–mass relation tells us this very clear fact, but this allows for a simple means of escape and deep reconciliation. As postulated above dark matter would repel itself, and universally photons comprise the micro-complexities of the subatomic, matter or dark matter. So yes, dark matter would repel all forms of motion (which we have already shown tautologically), even the very frequency of atomic resonance animating dark matter/energy.


## Strong Like Force

We will denote this dark matter field $\psi$. Exactly the same concept as "strong force," with respect to the concentrated nuclear force welding subatomic particles into existence, the force of $\psi = fx$. This force would exert three-dimensionally on each atomic mechanism as mentioned above. The force of the collective dark matter field would "attack" the integrity of each individual atom from every point available on the surface area. Again traveling at light speed. Not a collective linear force; the force would propagate to the center of each atom. The atomic mass of the matter in question creates the calculus of force with base $fh \Rightarrow fx$. $\psi$ would be stable in existence. The particle's magnetic spin would be both (-,+). Whichever charge would extend toward $\psi$ (creating a priority photon array emission), the like charge would manifest, creating the force of repulsion $\psi = fx$. Simply put, the dominant photon arrays emitted by visible matter(namely protons, considering the mass difference with respect to electrons) 'flips' a particular 'positive' or 'negative' quantum emission switch within dark matter's atomic structure, which constantly repels said matter, with opposing photon arrays. The opposing photon array is respectively dispersed, though the 'priority' emission is disproportionately generated. Thus the presence of mass would bend the composition of spacetime (dark matter) proportionate to the volume of repulsion (spherical obviously) with respect to the mass of the celestial body in question.

And more specifically not just contortion of the composition of spacetime around any particular massive manifold, but around each atom individually. This is in essence the engine of gravitation, the constant force of repulsion between matter and dark matter. The inevitable means to the end of course; it's complete equilibrium in the way of static resonance. Essentially the adhesive (dark matter) is eroding, as the proposed work implies. As the amalgam of dark energy continues to cool, an accelerated propulsion will continue to mount. As more atoms of dark matter awaken, the force of repulsion will grow and continually destabilize the symbiotic adjoining of matter and dark matter. A concept not yet covered simply addresses the idea of the cradle of dark matter affecting the velocity of celestial bodies by means of repulsion. This theory is built perfectly for such questions. Any point of repulsion on the surface area of our hypothetical body A would nullify any other point, and so on.

This field $\psi$ would not affect the inertia of the body in question due to equal force at each point {x, y, z}. $\psi$ would of course have mass; otherwise the particle would move at the speed of light. A logical assumption would be near the mass of a proton, though in essence would be a proton and electron simultaneously. Theoretically $\psi$ could repel both a proton and electron simultaneously, though even if that was not the case, the repulsion of the proton and $\psi$ would supersede electron and $\psi$. Thus still resulting in repulsion. Also mentioned the dark matter particle is in a constant state of binary duality. Specifically that of a complex integer. As mentioned above, existing on two planes would allow constant repulsion; the Euclidean field would continuously "fold" over, disallowing continuity, adding continued resistance. In simpler terms imagine trying to slide a piece of parchment into an envelope. Though every time you do, someone folds the envelope in half. The parchment would simply come in contact with the envelope and remain occupying its own space. The same concept applies to the subatomic. Consequently the dark matter particle is in a constant state of reversal, the channels of photon array flow are constantly altering its "present" direction.

**Fig 10**

Representation of an individual atomic mechanism comprising $\psi \Rightarrow$

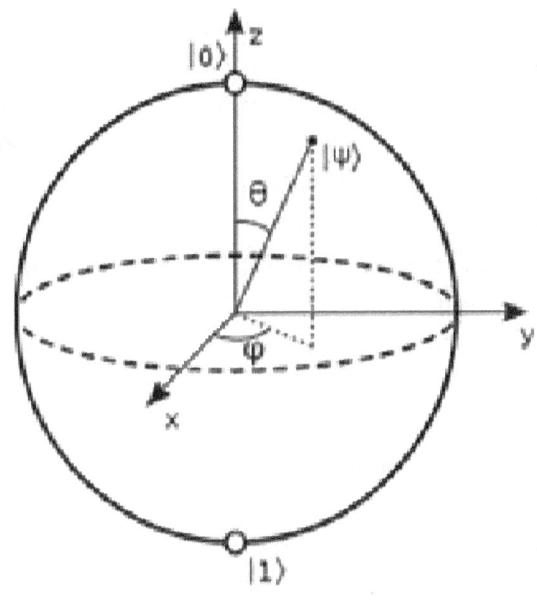


*Simultaneous electromagnetic spin of (+,-)
*The diagram details the Euclidean complex plane, spherical and thus three-dimensional. Individually each infinitesimal particle comprising dark matter is an essential qubit suspended within the complex plane.

A qubit by definition is an infinitesimal memory location. Qubits are arranged in superpositions, in which the given state is in a linear combination of state 0 and state 1. Just as the qubit assigns a state of (1,0), the atomic mechanism of dark matter continually deforms the photon array channeling emitted by the particle in question. The qubit has been popularized as the essential tool within the core programming of quantum computers, which propels the fantastic computational rate. In the same way that extraordinary potential resides within the silicon-based qubit, the material qubit, which is this work's aim of conveyance in describing the proposed makeup of dark matter holding enormous amounts of energy. A polymorphic state of existence, with no actual form or clear identity, the qubit or infinitesimal memory location of dark matter holds the lonesome responsibility of entire theoretical isolation. In a deep sense, dark matter holds no other "aim" than to stomp out energy exchange and motion of all variants. Visible matter is a particular mutation among the majority of the universe. Hyper Interactivity is very clearly not interwoven into dark matter's proverbial DNA.

The concept of each atomic mechanism, in a particular celestial structure, contributing a proportionate amount of energy contradicts relativistic explorations of gravitation. Within the core concept of relativity, the curvature of spacetime around mass/energy causes the pathing of a close object (satellite) to curve the original vector and time while the object begins revolving around the more massive celestial body.

Yes, the mass is proportional to the magnitude of curvature, though the spacetime *itself* curves the vector. Thus the most brutally complex point of relativity is that the mass of supermassive bodies themselves deforms spacetime, and the effect of gravitation is incidental to the effect of attraction itself. The algorithm presented above shows clearly that, by means of *the product with the average atomic mass*, energy allocated from each atomic mechanism varies proportionate to the unique atomic mass. This fact concerning the induction of the average atomic mass is very interesting for the sole purpose of demonstrating this identity clearly mathematically correlates with data concerning the composition of the celestial body in question. While employing combinatorial methods of arithmetic, the notice of each coefficient being that of the average atomic mass acted as a tool of orientation. The more accurate the results, the more accurately the average atomic mass would be applied to the coefficients in each operation.

## Anti Electromagnetism

Electromagnetism is the force in which an energy field envelops every electron (within a particular structure) in a particular atomic arrangement; this atomic arrangement charges the structure, resulting in a net force of attraction. This force creates an attraction to other structures in the field. By means of a net field, the collective energy rearranges the atomic structure and charges the structure to attract only particles of opposite charge. As stated above the dark matter/energy and visible matter photon array repulsion result in the gravitational energy allocated to each atom of celestial bodies in a net-like force similar to electromagnetism. The mathematical denotation presented above in Fig 9 quantifies this said gravitational energy into an identity expressed in newtons. As explained above $\psi$ field would apply an energy $fx$ to each atom (and in return an equal repulsion). Bending around each the net force would "flex" the atom and cause an energy cloud of photon dispersals to manifest. The same principle would apply, and a net field would propagate. Though because the dark matter particles stimulating the energy induction are perpetually repelled to matter, the force charges the atoms collectively (in a very weakened state) to attract "the same energy signature," including light. This proposition is the alternate view of the test conducted in 1919. As the results proved Einstein a genius at the time, obvious problems have been addressed in the proclamation of this clearly oppositional work. A logical question attacking the integrity of the core of this work's synthesis would pose, "Why are protons repelled by protons, electrons repelled by electrons, etc., if there is a supposed weak likeness attraction, namely gravitation?"

The answer is the repellant force of like particles is strong, though it does not "stack collectively" and thus does not create a "net force." As mentioned above electromagnetic fields collectively acquire energy from each atomic mechanism. The net force stimulates the field itself. Thus individually subatomic particles of visible matter will repel. Though as the proposed mass in question gains in magnitude, the gravitational attraction or "net weak like force" supersedes the individual repulsion. Subatomically it would serve logically to assume particles charged by simultaneously charged particles and magnetizing said particles results in attraction to particles with opposing magnetic photon array dispersal. Thus being "charged" by an $\psi$ field (simultaneously ±) would result in "like" attraction of the same photon emission signature as the atom itself (including light). Though only on a massive net scale explaining light vector deformation.

Dark energy would obviously be the photon array emissions of the dark matter itself, permeating all matter composed within each galaxy. The gravitational net energy stimulated by the presence of dark matter/energy enveloping visible matter. Just as the dark matter simultaneously emits positive and negative photon arrays (as shown above), "likeness attraction" stimulated by dark matter and visible matter interacting would be photon array dispersals that act much like alternating current or AC emissions. Alternating current is a type of electricity transfer that involves the generated power "cycling" through periods of linear travel, then reversing toward the source of emission. In this same way, likeness photon arrays would of course repel each other, though by process of photon array directional conjugates (alternating photon vectors). The opposing photon arrays of, for example, Earth and moon would "push" and "pull" on one another while cycling in a linear motion and momentarily reverting to the celestial body of emission. This photon array directional conjugates or "reversal photon vectors" would also simultaneously change the gravitational electromagnetic charge (+,-) in the same way as the dark matter/energy photon arrays of repulsion. Though instead of being characteristically repelled to all matter, like dark matter, the photon arrays stimulated by dark matter/energy and visible matter repulsion would reverse their vectors in cycles. The likeness photon arrays of gravitation would synchronize with surrounding gravitation fields as well, creating a harmonious dance of photon array dispersal and thus gravitational waves. This would allow for universal likeness attraction of celestial bodies.

The dark matter itself, cradling and enveloping visible matter, cosmologically baptizes the matter to simulate likeness-based attraction. The claim is essentially proposing that a celestial manifold should in all logical conception repel another manifold. The likeness repulsion should diverge the path of any manifold from another. Though through the intersection of the elusive dark matter/energy, there is the birth of the driving force of gravitation. Much like the concept of "strong force" mentioned above, this is also a force forged in tension and constant struggle that is responsible for nearly the entirety of motion in the universe. This natural mechanism of variance pulses and evolves not unlike the words of Descartes, "The universe is essentially a machine, not unlike a clock." And as such epochs of change shape the direction of the future.

## Predictions Made by This Model

The model of unfathomable mass of " Dark matter" packed in around galaxies (which is what this paper is describing) creating a repellant "stress-based" force, coupled with a collective spherical force bending the very composition of every individual atom, is predicted by this very explanation of gravitation. This "packing material"-like effect predicts certain characteristics already known of the universe concerning the existence of dark matter itself, the expansion of the universe, etc. As mentioned earlier *dark energy and dark matter* compose an estimated ninety percent of the matter in the universe. With the magnitude of energy emission necessary to power such a cosmological engine, this model would predict the existence of isolated concentrations of mass/energy "packed" in around galaxies out of obvious necessity. The enormous proximity between the visible universe and the majority of dark matter would also be explained by the force of repulsion and long-expressed intervals of time. With respect to the initial emergence of visible, interactive matter, in all probability this would have caused an insurmountable instantaneous moment of rapid apparent evacuation of the surrounding dark matter/ energy.

Though as proposed earlier, dark matter forged much later than visible matter. Again allowing the atomic structure we know today to form without the force of gravitation. Of course void of the continuous struggle of repulsion, matter cooled and spread in an unfettered and free way. Though the effects of gravitation soon became evident with the emergence of dark matter. This brings the argument to the structural erosion of the massive individual concentrations of antimatter. Imagine such a massive plume of raw material and, interestingly, the mass itself dwarfs our galaxy and all galaxies. While cradling the galaxies, much like a compass used in shipping navigation, the compass is suspended in a cradle that moves along a gyro-type mechanism—ignoring variant motion—and always remains on its own plane. The encompassing mass of dark matter that cradles the galaxies theoretically could be in the process of diverging the make-up layers of superposition at an accelerated rate. As mentioned this would explain the continuous expansion of the cosmos. The collective self-repellant nature of the antimatter qubit would, over a long enough time interval, diverge and "peel" outward into empty space.

Imagining the concentration and realistic geometry of such an amalgam plume of mass/energy, this work would predict the continuous shifting and constant divergence of the atomic antimatter composition in question. Like sand in the tide, portions would be stripped away, eroded, and pushed farther into empty space. And so on and so on without interruption. Over the consideration of a near-incalculable interval, a more obvious and extreme topological deformation of the $\psi$ field can be visualized. Eventual complete divergence of the antimatter field would result in an instantaneous vacuum effect. This would predict not just the expansion of the universe, but the acceleration as well. A stunning result that is clearly harmonious with the concepts of modern physics. The acceleration would be a result of the "outermost" layering of the dark matter field, having no further reinforcement to further diverge the collective repulsion energy waves being pushed into open space. As the calculus of layering strips further, less available field remains. Causing incremental energy loss of the field, which invites easier and faster layer stripping, and so on.

## Verifiable Data

The method this particular algorithm follows is a rigid though simple/tedious means of arithmetic operators. Avogadro's constant plays a particular role in this calculation in the way of providing the approximate number of atomic mechanisms within the body in question. The value of the constant being 6.02214076 x 10^23, which is a dimensionless value describing the number of atoms in twelve grams of Carbon-12, denoted a "mole." Average atomic mass must be calculated in order for the product to have a counterpart. For example Uranus is composed mainly of liquid/frozen water and methane. The ratio is respectively 9:1, so the average can be based on that of water, which has a near-exact atomic mass of 18 u. The unit of measurement is always performed in kilograms, thus taking one thousand grams (kilogram) divided by the atomic mass of water, 18, yields approximately 55. The product of 55 and 6.02214076 x 10^23 equals approx. 3.3 x 10^25.

This value, which is the approximated count of atoms composed within Uranus, is then multiplied by the identity shown above. The approximate value of the identity being 1.1091 x 10^-37, expressed in newtons, dictates the predicted allocation of likeness attraction emitted by one atomic mechanism. Yielding, as mentioned before, a net-like force exactly like that of the phenomenon describing electromagnetism.

**Fig 11**

[ref  https://www.hou.usra.edu/meetings/geodyn2015/pdf/5001.pdf]

***Solar mass: 1.98 x 10^30 kg***

Atoms comprised within the sun: 1.19 x 10^57

Approximate average atomic mass: 1.005 u

*Gravitational identity yields:*

*1000/1.005 = 995*

*995 x (6.022 x 10^23) x (1.98 x 10^30)*

*x identity  = 1.325 kg/m/s*

*Using Gravitational constant (6.67 x 10^-11) yields:*

*1.3215 x 10^20 kg/m/s*

### *Mercury mass: 3.3 x 10^23 kg*

Atoms comprised within Mercury: 5.55 x 10^48

Approximate average atomic mass: 35.8 u

*Gravitational identity yields:*

*1000/35.8= 27.93*

*27.93 x (6.022 x 10^23) x (3.3 x 10^23)*

*x identity  = 2.2 x 10^13 kg/m/s*

*Using Gravitational constant (6.67 x 10^-11) yields:*

*2.2 x 10^13*

### *Venus mass: 4.86 x 10^24 kg*

Atoms comprised within Venus:

Approximate average atomic mass: 25.8 u

*Gravitational identity yields:*

*1000/25.8 = 995*

*995 x (6.022 x 10^23) x (4.86 x 10^24)*

*x identity = 3.244 x 10^14 kg/m/s*

*Using Gravitational constant (6.67 x 10^-11) yields:*

*3.2435 x 10^14 kg/m/s*

### *Earth mass: 5.97 x 10^24 kg*

Atoms comprised within Earth: 1.356 x 10^50

Approximate average atomic mass: 26.5 u

*Gravitational identity yields:*

*1000/1.005 = 995*

*995 x (6.022 x 10^23) x (5.97 x 10^240)*

*x identity = 3.985 x 10^14 kg/m/s*

*Using Gravitational constant (6.67 x 10^-11) yields:*
*3.984 x 10^14 kg/m/s*


## Lunar mass: 7.35 x 10^22 kg

Atoms comprised within the moon: 2.0302 x 10^48
Approximate average atomic mass: 21.8 u
*Gravitational identity yields:*
*1000/21.8 = 48.65*
*48.65 x (6.022 x 10^23) x (7.35 x 10^22)*
*x identity = 4.905 x 10^12 kg/m/s*
*Using Gravitational constant (6.67 x 10^-11) yields:*
*4.905 x 10^12 kg/m/s*


## Vesta mass: 2.59 x 10^20 kg

Atoms comprised within Vesta: 6.44 x 10^45
Approximate average atomic mass: 24.2 (±1) u
*Gravitational identity yields:*
*1000/24.2= 41.32*
*41.32 x (6.022 x 10^23) x (2.59x 10^20)*
*x identity = 1.729 x 10^10 kg/m/s*
*Using Gravitational constant (6.67 x 10^-11) yields:*
*1.728 x 10^10 kg/m/s*


## Martian mass: 6.4 x 10^23 kg

Atoms comprised within Mars: 1.529 x 10^49
Approximate average atomic mass: 25.2 u
*Gravitational identity yields:*
*1000/25.2= 39.68*

*39.68 x (6.022 x 10^23) x (6.4 x 10^23)*
*x identity = 4.27 x 10^13 kg/m/s*
*Using Gravitational constant (6.67 x 10^-11) yields:*
*4.27 x 10^13 kg/m/s*

### Jupiter mass: 1.898 x 10^27 kg

Atoms comprised within Jupiter: 1.137 x 10^54
Approximate average atomic mass: 1.005 u
*Gravitational identity yields:*
*1000/1.005 = 995*
*995 x (6.022 x 10^23) x (1.898 x 10^27)*
*x identity = 1.267 x 10^17 kg/m/s*
*Using Gravitational constant (6.67 x 10^-11) yields:*
*1.267 x 10^17 kg/m/s*

### Saturn mass: 5.69 x 10^26 kg

Atoms comprised within Saturn: 3.409 x 10^53
Approximate average atomic mass: 1.005 u
*Gravitational identity yields:*
*1000/1.005 = 995*
*995 x (6.022 x 10^23) x (5.69 x 10^26)*
*x identity = 3.798 x 10^16 kg/m/s*
*Using Gravitational constant (6.67 x 10^-11) yields:*
*3.797 x 10^16 kg/m/s*

### Uranus mass: 8.68 x 10^26 kg

Atoms comprised within Uranus: 2.90 x 10^52
Approximate average atomic mass: 18.005 u

*Gravitational identity yields:*
*1000/18 = 55.5*
*55x (6.022 x 10^23) x (8.68 x 10^26)*
*x identity= 5.7914 x 10^16 kg/m/s*
*Using Gravitational constant (6.67 x 10^-11) yields:*
*5.793 x 10^16 kg/m/s*

### Neptune mass: 1.024 x 10^26 kg

Atoms comprised within Neptune: 3.342 x 10^51
Approximate average atomic mass: 18.005 u
*Gravitational identity yields:*
*1000/18= 55.5*
*55.5 x (6.022 x 10^23) x (1.024 x 10^26)*
*x identity = 6.83 x 10^15 kg/m/s*
*Using Gravitational constant (6.67 x 10^-11) yields:*
*6.83 x 10^15 kg/m/s*

### Pluto mass: 1.3 x 10^22 kg

Atoms comprised within Pluto: 7.88 x 10^48
Approximate average atomic mass: 46 u
*Gravitational identity yields:*
*1000/46 = 21.7*
*21.7 x (6.022 x 10^23) x (1.3 x 10^22)*
*x identity = 8.67 x 10^11 kg/m/s*
*Using Gravitational constant (6.67 x 10^-11) yields:*
*8.67 x 10^11 kg/m/s*

The overwhelming accuracy of these results gives an obviously clear glance into the true gravitational identity. Not an identity built on pure mathematical measurement, but one built on the conception of a mechanism. An engine of attraction, a powerhouse of gargantuan photon array resonance. And of course, as stressed, the conceptional identity must have an indelible abridgment to the effect being described. Namely gravitation.

## Summation

An amusing and constructive form of closure is a symbolic folk tale. One that is suitable for the particular genre of argument entails the short life of a group of Neanderthals. Assuming a moderately advanced version of speech existed among the members of said society. Imagine roles for each member much like today. Upon a period of great travel, a group of primitive men came upon a being unseen by man (until this moment). The animal had a sleek pointed face, colored and reflective fur-like coating, and small twig-like legs. And this animal had an incredible means of transport over treacherous water rapids—the members of the primitive society had noticed day to day the animal being on one particular side, then the other side of the river the next day. Though no one had seen the animal actually physically bring itself to cross the river. Og, the group philosopher and naturalist, was puzzled by the means by which the animal negotiated itself to the other side of the deadly river. The awe in the roar of the crowd stirred the inconsolable hunger to explain this phenomenon.

Og consulted the spirit realm through ritualistic activity. By the following morning, Og had received a vision of inspiration. Og called all the members to the sight of the mysterious creature. He announced his vision from the night prior in vivid detail. "When the night falls, the animal crawls across the stars because they are lower. Which is why the stars are visible only throughout the night, and we never see the creature cross." In one moment Og had solved a phenomenon that baffled the society, two birds and one stone (no pun intended). The moral of course being understanding and perception are typically non conducive to one another, without a divine prism in which to gaze.

This is a humorous example of the danger in which one is quick to correlate apparent characteristics with immature data sets. Very erroneous concepts can ripple from such arrogance and blatant naivety. The claim that general relativity makes concerning two bodies only influencing the "spacetime" both bodies reside in—which in turn creates the motion we know as gravitation—is a complete oversimplification. This work clearly believes the mechanism is powered by means of "curvature" or topological deformation. Though this work entirely disagrees with the means of deformation, general relativity claims.

The proposed idea presented here offers a clear and simplistic (not to mean natural) means of understanding cosmic deformation in contrast to an unexplainable source of potentially infinite power (spacetime) by means of drawing a deep connection using sources that have been verified to exist. Data, in short, is the compass of this particular adventure. The concept of matter–antimatter interaction was presented with detailed outlines of the process weighed against potential means of explanation. Determined to be the most reliably consistent tool for explanation. This process was designed with the quantum in full control. Planck's constant is the epicenter of the identity, the constant that describes the subatomic. Using this identity trials were conducted, observing the (if any) deviation of particular test subjects.

Mathematical adjustments were painstakingly made, attempting to refine the decimal convergence to a higher degree of accuracy. This is the process of forging understanding from a purely naturalistic prism. Rigor and perseverance with essential tools, though the time put in "crunching" the numbers, is the only real currency of truth. This is the process of interpretation of synthesis.

# References

https://www.physlink.com/Education/Askexperts/ae198.cfm#:~:text=Einstein%20predicted%20that%20light%20should%20be%20bent%20by,had%20been%20bent%20exactly%20as%20Einstein%20had%20predicted

https://www.britannica.com/science/binding-energy

https://www.britannica.com/science/electromagnetism

https://www.space.com/universe-expansion-rate-hubble-telescope-measurements

https://www.hou.usra.edu/meetings/geodyn2015/pdf/5001.pdf

# CONCLUSION

This work was motivated by the concept of logical explanation and academic dissent. The meaning of knowledge is compounded by two differing concepts, that of experience and intelligence, both of which share the sole compatible characteristic of depth. Depth itself is three-dimensional, which is the type of vision one must have to not only ask the questions of utmost importance, but to possess the rigor and tenacity to prolong the bout of synthesis long enough to stimulate results of interest and categorically isolate the patterns that inevitably emerge in endomorphism.

Although the three parts of this work presented seemingly disconnected databases of information, all three concepts are resonating dimensions of the greater universal machine. Part One begins with a general description of the "default" planetary genetics concept concerning planetary spacing. Exposing a template of construction with respect to the exoplanet spacing, along with the emergence of a completely original numeric series that shared a connection with the Fibonacci sequence and bringing clear insight and shadowing to the widely accepted Titus–Bode law. Part Two becomes more specific, clearly demonstrating the numeric pattern with respect to planetary dynamics and again the Fibonacci convergence, while allowing for a simplistic and accurate means of using a spectrum of value to determine an exoplanet's habitability. Further demonstrating the depth of the connection concerning natural mechanisms and the Fibonacci sequence.

And finally Part Three describes the universal mechanism of gravitation that is the ultimate driving force of Part One and Two, responsible for nearly all the motion in the universe. So there is a clear succession of understanding or logistic un-Earthing (if you will).

Numeric translation of the physical universe is the only means of compromise we as humankind can offer future generations as a means of interpretation, faced with the mounding data being processed at accelerated rates. The calculus of change can be described in any mathematical system, from general population trends to quantum dissection within accelerators. As this work proposes, those same subtle numeric patterns responsible for general variance exist within the cosmological sheath of understanding for centuries without coherent explanation. Simplistic results, along with the highest level of conception, comprise the means of producing explanations well beyond the allowable elevation of exploration.

# ABOUT THE AUTHOR

---

Bruce Robert Nye, Jr. is a self-professed autodidact who has taught himself advanced concepts to relate the perception of reality to empirical constructs of logic. He is currently operating a chain of commercial coin operated laundromats alongside his brother and parents. A family man, he lives in southern Florida with his wife and three children.